MEGALODON!

The Complete History Of The Largest Predatory Shark That Ever Lived!

MEGALODON!

The Complete History Of The Largest Predatory Shark That Ever Lived!

Robert L. Fuqua

Copyright © in 2022 by Robert L. Fuqua
All Rights Reserved

Author's Cataloging In Publication Data
Fuqua, Robert L.
MEGALODON! The Complete History Of The Largest Predatory Shark That Ever Lived!
1. Megalodon
2. Shark teeth
3. Paleontology
4. Fossils
ISBN: 9798361311347

Table of Contents

Chapter	Page
1. MEGALODON!	1
2. Fossilization	5
3. Megalodon Teeth	7
4. Evolution Of Sharks	25
5. Evolution Of The Megalodon	29
6. The Megalodon Body	47
7. Megalodon Sensory And Internal Organs	53
8. Where Megalodon Teeth Have Been Found	61
About The Author	75

List of Figures

Figure	Page
1. 5.75-inch (14.6 cm) Megalodon tooth.	viii
2. A perfect 4.75-inch (12.06 cm) Megalodon tooth.	4
3. Fossil shark tooth colors.	6
4. Display of various fossilized shark teeth.	8
5. Developing shark teeth.	10
6. Great white shark jaw with a symphyseal tooth.	11
7. Upper cutting and lower grasping Megalodon teeth.	14
8. Pathological deformity of a lower Megalodon tooth.	15
9. Tooth arrangement of a Megalodon shark.	17
10. Upper and lower anterior Megalodon teeth.	18
11. Upper intermediate Megalodon tooth.	19
12. Upper and lower lateral Megalodon teeth.	20
13. Upper and lower posterior Megalodon teeth.	20
14. The geologic time scale.	26
15. A 4 inch (10.2 cm) Megalodon tooth.	28
16. Evolution of the great white shark.	32
17. *Palaeocarcharias stromeri*.	33
18. *Cretalamna appendiculata* tooth.	34
19. *Otodus obliquus* tooth.	35
20. *Otodus aksuaticus* tooth.	36
21. *Carcharocles auriculatus* tooth.	37
22. *Carcharocles angustidens* tooth.	38
23. *Carcharocles chubutensis* tooth (back and front).	39-40
24. *Carcharocles Megalodon* tooth (back and front)	41-42
25. Evolution of the *Carcharocles Megalodon*.	44
26. Comparison of a Megalodon and great white tooth.	46
27. General shape and fins of a Megalodon.	47
28. Megalodon's cartilage skeleton.	49
29. Fossilized *Otodus Obliquus* centra.	49
30. Megalodon skull and jaws.	51
31. Drawing of Megalodon brain.	54
32. Ampullae of Lorenzini.	57
33. Earth 390 mya.	64
34. Earth 152 mya.	65
35. Earth 94 mya.	66

36. Earth 50 mya.	67
37. Earth 14 mya.	67
38. A 4 inch (12.1 cm) Megalodon tooth.	73
39. Photo of the author.	74

List of Tables

Table	Page
1. Chemicals that contribute to a fossil's color.	6
2. Explanation of the Megalodon dental formula.	16
3. Ancestors of the *Carcharocles Megalodon*.	43
Table 3 repeated	63

Credits

All photographs and drawings are by the author unless credited below.

Developing shark teeth image on page 10 is in the public domain.

Shark jaw with a symphyseal tooth photograph on page 11 is in the public domain.

Palaeocarcharias stromeri photograph on page 33 is by Ghedoghedo and licensed under the Creative Commons Attribution-Share Alike 4.0 International license.

General shape and fins of the Megalodon image on page 47 is by Dinosaur Zoo and is licensed under the Creative Commons Attribution-Share Alike 3.0 Unported license.

Megalodon's cartilage skeleton image on page 49 is by Albert C. L. G. Gunther and is in the public domain.

Ampullae of Lorenzini image on page 58 is by Bernard Dupont and is licensed under the Creative Commons Attribution-Share Alike Generic (CC BY-SA 2.0).

Paleogeographic maps used with permission of C.R. Scotese, PALEOMAP Project. Scotese, C.R., 2001. Atlas of Earth History, Volume 1, Paleogeography, PALEOMAP Project, Arlington, Texas, 52 pp.
Figures: 33, 34, 35, 36, 37

Photo of the author on page 74 is by Linda Donovan.

Other Books By Robert L. Fuqua

A Living Jewel
A Beginner's Guide To Saltwater Aquariums
Published 1995, Second Edition 2009

Hunting Fossil Shark Teeth In Venice, Florida
The Complete Guide: On The Beach, SCUBA, and Inland
Published 2011

Fossil Shark Teeth Of Venice, Florida
The Paleogeology Behind How Shark Teeth, And Other Fossils, Ended Up In The Venice Area
Published 2017

Algae The Alligator
The Alligator That Was Raised By Turtles
Published 2017

A Brief History Of The Earth
Formation Of The Earth, Plate Tectonics, Volcanoes, Supercontinents, Sea Levels, Ice Ages, Climate And Life
Published 2019

Florida Fossil Shark Teeth Identification Guide
The Fossil Shark Teeth Most Commonly Found In Florida
Published 2020

Could You Have A Gluten Problem?
A Concise Guide To Understanding, And Living With, Gluten Sensitivity And Celiac Disease
Published 2020

Figure 1. 5.75 inch (14.6 cm) Megalodon tooth.

Chapter 1

MEGALODON!

The image on the cover of this book, and Figure 1 on the opposite page, is a 5.75 inch (14.6 cm) Megalodon tooth that I found SCUBA diving off Venice, Florida in June 2016. Tropical storm Colin had just blown through and the wave action exposed this tooth in 23 feet (7 meters) of water. I have found many Megalodon teeth while SCUBA diving and this was by far my biggest and best ever.

The Megalodon was the largest predatory shark that ever lived and the largest of them were probably a little over 60 feet long (18.3 meters). The tooth on the cover was from a Megalodon that was about 57 feet long (17.4 meters). The rule of thumb is that there will be 10 feet (3.05 meters) of Megalodon for every inch (2.54 centimeters) of tooth. By comparison, the largest modern predatory shark is the great white, which can reach a length of 20 feet (6.1 meters). Other modern sharks, such as the whale shark, Greenland shark, and the basking shark, can also get very large, but those sharks are filter feeders that consume plankton and small fish. They are not a threat to humans.

Some common sharks that we see near the beach include bulls, lemons, and sandbar sharks. Typically, these sharks are around six feet (1.8 meters) long, some shorter and some larger. But, when we see them most people are filled with fear. Others are filled with awe at seeing such an amazing creature. Occasionally, we will see larger sharks such as the tiger, mako, or the great white. The fear and awe levels rise. Now imagine seeing a Megalodon that is up to three times larger than the largest great white, with a dorsal fin seven feet (2.13 meters) tall. The fear and awe levels are now off the chart! But, there is no need to worry about going to the beach, the Megalodon has been extinct for 2.6 million years.

This book is going to discuss everything we know about the Megalodon, the most fascinating of all sharks. As will be discussed in more detail later, sharks do not have bones. Instead, their internal structure is made up of cartilage, which can fossilize, but only under very special conditions. A very limited number of fossilized shark vertebrae (centra) cartilage have been found and we learn everything we can from them. Shark teeth are made of harder material, and enameloid, that fossilizes easily and are much more commonly found. All of the teeth you will see in this book are fossilized and life size unless otherwise noted.

Most of what we know about Megalodons comes from the fossil record of their vertebrae centra and teeth. We can conclude many things from these items, but not everything we want to know. The next best thing we can do is to compare living sharks to the Megalodon fossils and extrapolate what the Megalodon must have been like. The closest living relative to the Megalodon is the great white shark. So, we generally project that the Megalodon looked like a very large great white, with all its various features. This is very reasonable because the great white and the Megalodon both belong to the Lamniformes order, which will be discussed in detail later. In this book I will detail if what we know is based on the fossil record or an extrapolation from the great white.

We are going to cover a lot of material in this book. So, it will be useful to give a quick overview of where we are going. The first thing that we need to do is have a quick overview of fossilization. Next, to put those fossilized Megalodon teeth in context, we need to understand the basic structure of those teeth and how they developed. You might be surprised at the amount of information there is regarding shark teeth.

Since Megalodons did not just appear out of thin air (water) we need to know something about the evolution of sharks in general. This will lead to a detailed discussion of the evolution of the Megalodon, its ancestors, and its close relatives. From there we move on to details about the body of the Megalodon, its size, shape, and structure. After that we will discuss the sensory systems and internal organs of the Megalodon. The sizes of some of these organs may be a surprise to the reader. Finally, we will have an overview of where Megalodon teeth have been found as well as a few other interesting details.

Figure 2. A perfect 4.75 inch (12.06 cm) Megalodon tooth.

Chapter 2

Fossilization

Figure 2 is a perfect 4.75 inch (12.06 cm) fossilized Megalodon tooth. Much of this book talks about fossil shark teeth. So, now is a good time for a quick overview of the fossilization process. The dictionary definition of a fossil is that it is the hardened remains or imprints of plant or animal life from some previous geologic period. That is a fine definition, but, let's take it several layers deeper. Most plants or animals will never fossilize. This is because several conditions must be just right for fossilization to occur. Generally speaking, the dead plant or animal must have hard parts such as bone, teeth or an exoskeleton. However, under the right conditions soft parts, such as cartilage, can fossilize.

The dead animal must quickly be buried in wet sediment. This sediment compacts around the remains and starts the next process. That process is for the mineral salts in the sediment to start to permeate the open spaces of the item at a microscopic level. Eventually the minerals will add weight to the item, change its color, and make it harder.

The color of a fossil depends on the minerals that permeate it from the sediment. Figure 3 shows several shark teeth that I found while SCUBA diving off Venice, Florida. The three larger teeth are Megalodons. The other three are a lemon shark and two bull sharks, going from left to right. You can see that they range from jet black, to bluish, brownish, and even reddish. The tooth in the center is grayish yellow with a darker tip. This is an example of an interrupted fossilization process. The whole tooth was fossilized to the gray-yellow color but then something interrupted the process and only the tip continued to fossilize to a darker color.

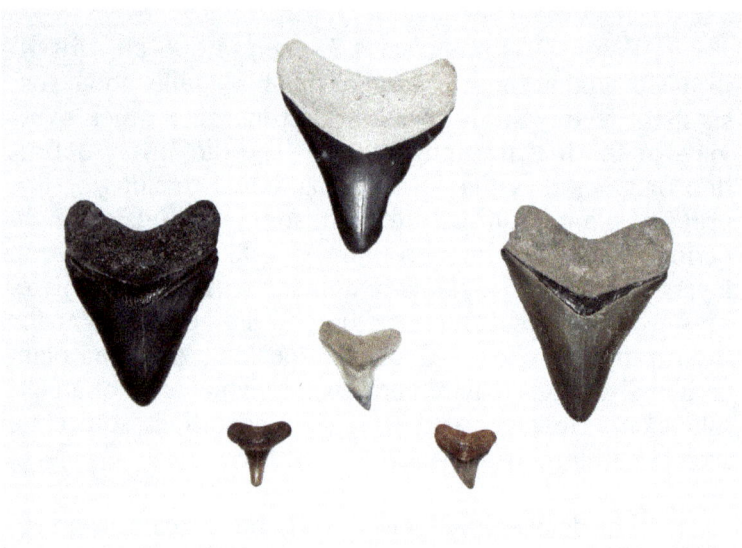

Figure 3. Fossil shark tooth colors.

    Phosphate in the sediment gives the fossil a black color. Iron gives the fossil a reddish-orange color. Limestone gives the fossil a grayish-yellow color and tannins give it a dark brown color. This information is summarized in Table 1.

Black	Phosphate
Reddish orange	Iron
Grayish yellow	Limestone
Brown	Tannins

Table 1. Chemicals that contribute to a fossil's color.

Chapter 3

Megalodon Teeth

What are the first physical characteristics you think of when you think about sharks? The first things that come to my mind are the dorsal fin and all those teeth. Since a shark's teeth are very important, and tell a lot about the shark, they deserve a whole chapter. We will discuss the fins in a later chapter. All the shark teeth that you will see throughout this book are fossilized and shown actual size, unless otherwise noted. I have given lots of lectures on fossils and fossil shark teeth and Figure 4 is a display I often took along. It is a good example of the wide variety of sizes and shapes that can be found in different shark species. They are placed on the acrylic board so that both sides of the teeth can be seen. The upper and lower teeth are properly positioned so that you can see the differences between them. The picture is about half the actual size. The two Megalodon teeth in the center are 2.75 inches (7 cm) long for the upper and 2.125 inches (5.4 cm) long for the lower. I wanted to show what they are like but did not want to use teeth so big that they overwhelmed the display. Also, I did not want to take a chance of damaging nicer, bigger, teeth as everyone looked at them.

The first part of this chapter will discuss the basic characteristic of all shark teeth. This is important because the basics that will be discussed also apply to Megalodon teeth. Then we will finish the chapter by discussing the details of Megalodon teeth. We will start this discussion of shark teeth by first looking at shark jaws.

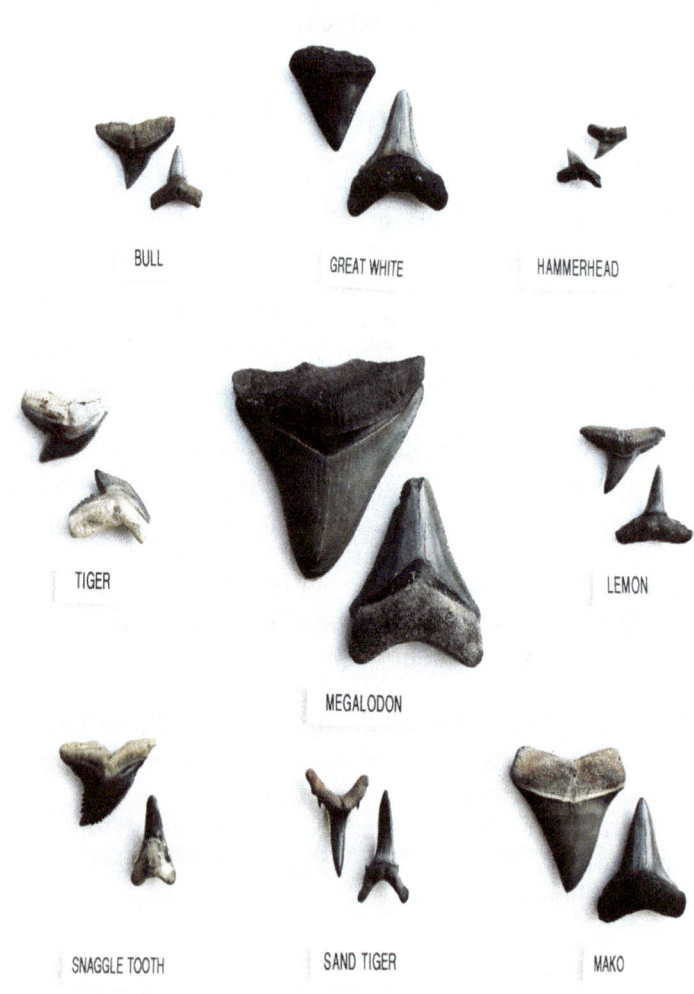

Figure 4. Display of various fossilized shark teeth.

Shark jaw anatomy

When we visualize a shark's jaw, we think of the sharp white teeth that are clearly visible at the very front of the top and bottom jaws. These are called the primary teeth and are the ones that do the biting and cutting. Those teeth are actually only a fraction of the teeth the shark has in its mouth at any one time.

As the shark bites its prey, and then possibly struggles with it, both the shark and the prey will be turning and twisting, one to get away and the other to consume a meal. This puts considerable strain on the shark's teeth and some of them can get broken or lost in the struggle. You may find a tooth with a broken tip or a small chunk missing from the side. This damage is generally caused by the shark biting into bone when it was feeding. Since a shark depends on a good set of sharp teeth for its survival, it cannot afford to lose too many teeth. To survive, the shark needs a constant supply of new, sharp teeth to replace the ones that have broken or fallen out. The lower Megalodon tooth in Figure 4 is an example of a tooth with a broken tip.

If you look at a shark's jaw from the inside perspective you will see numerous rows of teeth lying flat against the inside of the jaw, on both the top and bottom. Figure 5 shows a life-size view of the inside of a modern bull shark's upper jaw. The teeth nearest the primary teeth are almost ready for use and the ones deeper inside are still developing. The primary teeth are constantly moving forward until they fall out of the mouth. At the same time, the next rows of teeth in the jaw are constantly moving forward to replace the primary teeth. This process ensures that the shark is always well equipped with a full set of sharp teeth.

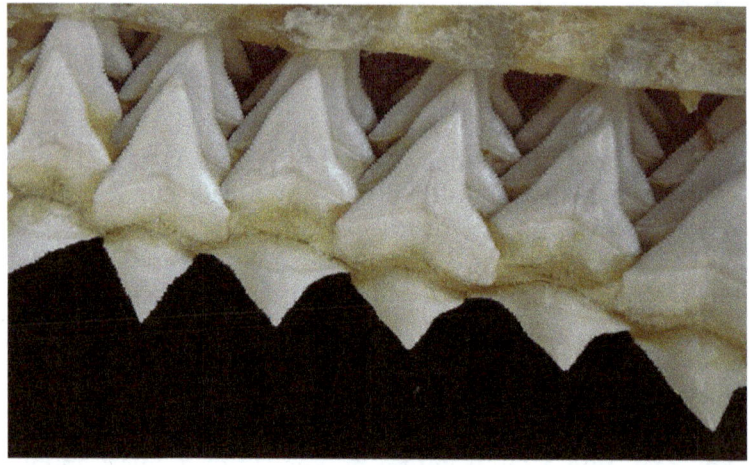

Figure 5. Developing shark teeth.

In addition to the rows of teeth, it is interesting to note the teeth that are sometimes found right in the midline where the left and right jaws join. These are called symphyseal teeth and they are considerably smaller than their companions. They also usually have a significantly different shape. Symphyseal teeth are rare and highly prized among fossil shark teeth collectors. Figure 6 shows an example of a small symphyseal tooth right on the midline of the jaw between two normal-sized front teeth. This picture is of a modern great white shark's upper jaw and is half the size of the actual jaw.

Shark tooth anatomy

Not all teeth in an individual shark's jaw are shaped the same. Their teeth can be very different front to back and top to bottom. This difference can be in size, shape or both. The shape of the teeth also changes with age, whether it was a primary or developing tooth, and sometimes based on the shark's sex. It is also true that the teeth from the same genus, but different species, of sharks can be very similar and often hard to tell apart.

The flatter side of the teeth face the front, ocean side of the jaw. The side of the tooth that is oriented toward the throat is rounder and more attractive. This is the side that we generally display when showing off our finds and we typically refer to this side as the back or display side. The flatter, front side slices into the prey and the curved back side starts forcing the flesh back into the mouth. This combination makes the tooth very efficient at removing flesh from the prey.

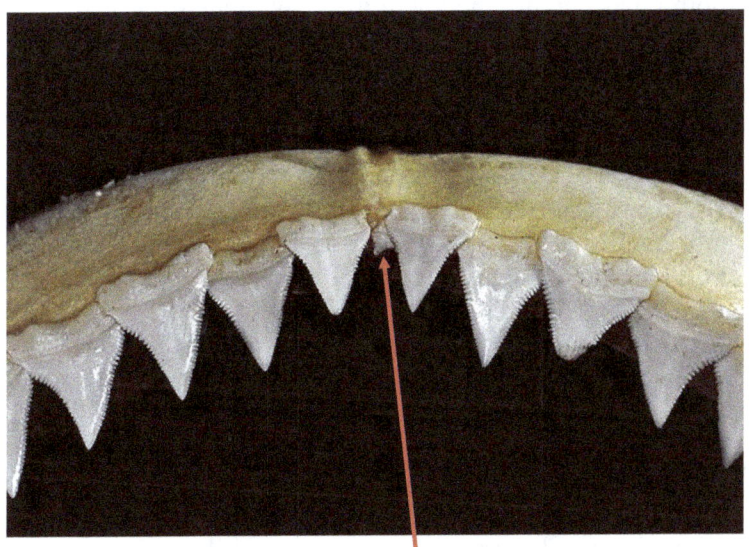

Figure 6. Great white shark jaw with a symphyseal tooth.

Beyond the size and shape of the teeth, there are many other characteristics and descriptive terms that need to be understood. These terms and characteristics are described below.

Anterior - the teeth in the front of the jaw.
Blade - sometimes called the crown - the exposed part of the tooth which is covered with enameloid.
Bourlette - sometimes called the dental band - this is the chevron-shaped band on the back of the tooth that separates the blade from the root. It is a distinct identifying feature on Megalodon teeth and some of its ancestors.
Cusp - small blades that exist on either side of the main blade.
Cutting teeth - have broad, sharp blades that are efficient at removing flesh.
Dental formula - a convenient method for recording the tooth types and numbers within a row of teeth.
Grasping teeth - are more stout, narrow, and pointed than the cutting teeth and are used to restrain prey in the mouth.
Lateral - the teeth on the sides of the jaw.
Notch - is a distinct change in the shape of the side of the blade.
Nutrient groove - is a small groove often surrounding the nutrient pore.
Nutrient pore - is a small hole in the root through which blood and the nerves entered the tooth.
Posterior - the teeth in the rear of the jaw.
Root - is the base of the tooth where it is attached to the jaw.
Root lobe - refers to the sides of the root and is normally used to describe roots with pronounced ends.
Serrations - very small, sharp, notches along the side of the blade - used to increase the cutting ability of a tooth.
Shoulder - the area of the side of the blade where it joins with the root.
Symphyseal teeth - small teeth that are sometimes found on the center line of the shark's jaw and often have a very different size and shape from those on the right or left sides of the jaw.

Other characteristics of fossil shark teeth

All shark teeth have a hyper mineralized shell on the blade and we call this shell enameloid. Inside this shell is the dentine core. The root consists of a bone-like material. The dentine core of all sharks, except the Laminformes order, is made of orthodentine. Teeth with orthodentine have a hollow pulp cavity surrounded by orthodentine and then the enameloid, all connected to the bone-like root. Laminformes teeth do not have that pulp cavity. The bone-like root extends upward inside the crown as osteodentine and is covered by enameloid. As we will see in Chapter 5, this difference between teeth with orthodentine and osteodentine is a major factor in the history of the Megalodon. As mentioned in Chapter 2, all fossils get their color based on the chemicals that were in sediment they rested in while mineralization was taking place. So, Mega-lodon teeth can be found with a wide variety of colors.

When measuring a shark tooth, we measure from the point of the blade to the end of the root lobe. Since the measurement to each lobe will be different, we use the longer of the two lengths when describing the tooth. This is a very important measurement when discussing Megalodon teeth.

Shark teeth can be categorized into two types: cutting and grasping. The upper teeth are flatter and have very sharp edges that are sometimes serrated. Sometimes the serrations are so fine that they are hard to see. In this case, run the end of a fingernail along the side of the blade. If it is serrated, you will feel those serrations. These are the cutting teeth. The lower teeth are not as wide, but are more stout, and sometimes serrated. These are the grasping teeth.

Cutting upper tooth
Grasping lower tooth

Figure 7. Upper cutting and lower grasping Megalodon teeth.

Perhaps the most rare of all are the teeth that have natural pathological deformities. These teeth have unusual twists, curves, or splits that are the result of disease or a healed injury that the shark suffered during its life. Figure 8 is an example of a pathological deformity of a 5-inch (12.7 cm) lower Megalodon tooth. The side of the crown of a normal Megalodon tooth is usually smoothly curved. This tooth has a very distinct change in that curvature as the side of the crown approaches the root. It angles down a bit about mid-way and then bends up toward the root. There is a partially healed crack on the display side where the blade starts angling down. Whatever caused that crack probably disrupted the growth of this tooth. The other side of the crown developed normally.

Crack

Figure 8. Pathological deformity of a lower Megalodon tooth.

Megalodon dental formula

The teeth in a shark's jaw are often described by a dental formula. The dental formula for the Megalodon is:

2.1.7.4
3.0.8.4
Megalodon dental formula

These formulas can be confusing until you learn the details of what they mean. Then it becomes very obvious. The numbers represent how many of various types of teeth are in a shark's mouth. The top line refers to the upper jaw starting at the midline and working back to the last tooth in that jaw. The same thing applies for the bottom line and the lower jaw. Table 2 explains the numbers in the Megalodon dental formula.

2 - two upper anterior (front) teeth
1 - one upper intermediate tooth
7 - seven upper lateral (side) teeth
4 - four upper posterior (back) teeth

3 - three large lower anterior teeth
0 - no lower intermediate tooth
8 - eight lower lateral teeth
4 - four lower posterior teeth

Table 2. Explanation of the Megalodon dental formula.

This formula represents the maximum number of lateral and posterior teeth. Actually, the Megalodon can have fewer lateral and posterior teeth in both upper and lower jaws, as we will see below regarding the great white shark. Figure 9 shows the full upper and lower tooth arrangements for a Megalodon. The teeth shown are an unassociated set from my collection. Unassociated means that the teeth were all found individually in many different places, which explains the variety of colors. An associated set would be found together in one place, which would be very rare.

In Figure 9 you can also see the various sizes and shapes of Megalodon teeth. The anterior teeth are the biggest and widest of all the teeth in that jaw. The upper intermediate teeth are thinner and much smaller than the anterior teeth and sometimes have a slight forward curve. The lower jaw does not have an intermediate tooth. The upper and lower laterals are of medium size and thinner than the anterior teeth. They may be slightly curved toward the rear. Posterior teeth are small and may also have a curve. Figures 10, 11, 12, and 13 show photographs of upper and lower anterior, intermediate, lateral, and posterior teeth respectively.

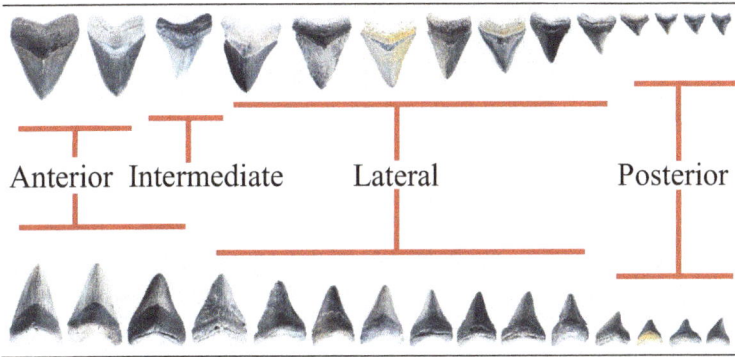

Figure 9. Tooth arrangement of a Megalodon shark.

Now that you have seen the Megalodon dental formula and the associated teeth, let's take a look at the great white dental formula.

$$\frac{2.1.7.4}{3.0.7.4}$$
Great white dental formula

Figure 10. Upper and lower anterior Megalodon teeth.

Figure 11. Upper intermediate Megalodon tooth.

Figure 12. Upper and lower lateral Megalodon teeth.

Figure 13. Upper and lower posterior Megalodon teeth.

This formula shows the full complement of teeth. However, the numbers of lateral teeth actually range from four to seven, in both the upper and lower jaws and the posterior teeth have a range of three to four, in both the upper and lower jaws. So, the following modification to the above formula provides that extra layer of detail.

2.1.4-7.3-4
3.0.4-7.3-4
My modified great white dental formula

Given all the similarities between the great white and the Megalodon shark, it is likely that the Megalodon also had a range of numbers of upper and lower, lateral and posterior teeth. So, the following modified Megalodon dental formula is more descriptive. It shows a range of four to seven for upper lateral, four to eight for lower lateral, and three to four for upper and lower posterior teeth.

2.1.4-7.3-4
3.0.4-8.3-4
My modified Megalodon dental formula

How many teeth?
The ramification of this is that the total number of teeth that any given Megalodon had depends on how many intermediate and posterior teeth it had. So, let's start to determine the maximum number of teeth a Megalodon could possibly have. The Megalodon had one row of primary and four rows of developing teeth that are almost ready to go into service. There is some speculation that the Megalodon had five rows of developing teeth. Of course, there will be additional layers of embryonic teeth even deeper in the jaw. But, for this book we will use the one row of pri-

mary teeth and four rows of developing teeth that are in the last stages of development.

On each side of the upper jaw there will be a maximum of two anterior, one intermediate, seven lateral, and four posterior teeth. This gives us 14 primary teeth on each side of the upper jaw for a maximum total of 28.

On each side of the lower jaw there will be a maximum of three anterior, zero intermediate, eight lateral, and four posterior teeth. This gives us 15 primary teeth on each side of the lower jaw for total of 30.

The combined total is a maximum of 58 upper and lower primary teeth. Now considering that there are five total rows, the maximum number of Megalodon teeth is 290.

Following the same logic, let's look at the minimum number of teeth. On each side of the upper jaw there will be two anterior, one intermediate, and a minimum of four lateral and three posterior teeth. This gives us 10 primary teeth on each side of the upper jaw for a total of 20.

On each side of the lower jaw there will be three anterior, zero intermediate, and a minimum of four lateral and three posterior teeth. This gives us 10 primary teeth on each side of the lower jaw for total of 20. The combined total is 40 primary teeth. Considering that there are five total rows, the minimum number of Megalodon teeth is 200.

So, the question of how many teeth the Megalodon had depends on the number of laterals and posteriors. The range becomes 200 to 290. This does not include any symphyseal teeth that might exist.

There is not a precise measurement of how many teeth any shark loses in a week, a year, or a lifetime. Estimates range from a few thousand to 20,000 teeth lost in a lifetime for a great white shark. So, let's do a little analysis to help determine what is reasonable for a Megalodon.

To start with, we need to examine how long a Megalodon's lifetime would be. Estimates range from 30 to 70 years. There have been several confirmations of a Megalodon's age at death of around 50 years by counting the growth rings in the vertebrae centra. So, for this analysis we will use a lifetime of 50 years.

If that Megalodon lost just one tooth per week, that would be two percent of its primary teeth each week. That would also mean 52 teeth per year, and 2,600 teeth over a 50-year lifetime. At this rate it would take 50 weeks to replace all of its primary teeth.

If a Megalodon lost one tooth per day, that would be seven teeth per week or 14 percent of its total primary teeth. That would also be a loss of 365 teeth per year and 18,250 teeth over a 50-year lifetime. At this rate it would take a little over seven weeks, to replace all of the primary teeth.

Now, what to make of these numbers? A large, active, predatory Megalodon eats large marine mammals. Marine mammals have lots of bones. Large whales have large bones that can cause lots of damage to Megalodon teeth. Taking a year to replace all 50 primary teeth would mean that the Megalodon would become a very inefficient eater and would probably starve. So, it seems that the number of lost teeth over a 50-year lifetime should realistically be around 15,000 to 20,000 teeth. If the Megalodon lived to 70 years old, there would be even more teeth lost.

Chapter 4

Evolution Of Sharks

We will start this discussion by quickly reviewing the formation of the Earth and the eventual development of life. The Earth, and the other planets, formed about 4.5 billion years ago from the accretion of material from the solar nebula.

The primitive Earth was very hot with lots of water. The water contained lots of chemicals from Earth's formation as well as expulsions from volcanos. The Earth's environment was basically a chemical stew slowly starting to create life. The earliest life on Earth was cyanobacteria that existed 3.7 billion years ago. Single-celled life formed and started to slowly evolve into simple multi-celled organisms. But, all that changed very rapidly 540 million years ago (mya) during the geologic period known as the Cambrian. It is often referred to as the Cambrian Explosion. It was not a big "boom" kind of explosion with things being destroyed. It was the kind of explosion where things changed and expanded rapidly. In this case, life finally took hold and those simple multi-cellular life forms started to rapidly grow and become more complex. During the Cambrian there were huge numbers of very unusual life forms. The Earth was full of biological niches that were rapidly exploited and filled by all this new life. For a quick reference of the geologic time scale, see Figure 14. For a more detailed explanation of these events, please read my book *"A Brief History Of The Earth."*

MEGALODON!

MYA	Era	Period	Epoch	Activity
0.0				Today
0.011			Holocene	
2.58		Quaternary	Pleistocene	
5.3			Pliocene	
23.0			Miocene	Sharks are abundant
33.9			Oligocene	
55.8			Eocene	
65.5	Cenozoic	Tertiary	Paleocene	Rise of mammals Dinosaur extinction
145.5			Cretaceous	
199.6			Jurassic	Modern sharks
252.2	Mesozoic		Triassic	
298.6			Permian	
359.2			Carboniferous	
416			Devonian	Primitive sharks
443			Silurian	
488.3			Ordovician	
542	Paleozoic		Cambrian	"Explosion" of life
2500	Proterozoic			
3800	Archean			First life - cyanobacteria
4500	Hadean			Formation of Earth

Figure 14. The geologic time scale.

First fish

Eventually many of those new and unusual life forms died off. This happened for many reasons. One was that the biological niche that they filled became a dead end and disappeared. Only life forms that could exploit numerous niches would continue to survive. Fish were a good example of this. The first fish were jawless. They had suckers and attached themselves to other creatures to gain nourishment. Eventually some fish started to develop jaws which allowed them to gain nourishment in a more efficient manner. Some of the jawed fish had bones. Other jawed fish had cartilage for their internal structure. The jawed fish with cartilage eventually became sharks.

Evolution Of Sharks

First sharks

Primitive sharks evolved around 420 mya. The cartilage in their bodies made them very flexible and maneuverable, and they were easily able to twist and turn while they chased their prey. They could also bite and eat their food with their jaws and rapidly evolving teeth. So, sharks became well adapted to their environment a very long time ago. Modern sharks appeared around 200 mya. The numbers of sharks continued to increase, and by 23 mya sharks had become abundant.

Today, after 420 million years of evolution, sharks have become very finely tuned and perfectly adapted for life in the ocean. In the next chapter we will continue this discussion as evolution continues and eventually leads to the Megalodon.

Figure 15. A 4 inch (10.2 cm) Megalodon tooth.

Chapter 5

Evolution Of The Megalodon

Scientists use a system called taxonomy to classify all living things. You are probably somewhat familiar with this system which divides life into eight rankings. These rankings range from the most general to the most specific in this order: domain, kingdom, phylum, class, order, family, genus, and species. It is a brilliant system which helps a person to know exactly what organism someone is referring to, even if they were talking to someone in a different country. For example, *Carcharhinus plumbeus* is the genus and species for the sandbar shark, and *Carcharhinus leucas* is the genus and species for the bull shark. They are very closely related, but different from one another. An organism gets classified to one of these ranks for many different reasons, to include physical similarities or differences, location where they lived, how many years ago the specimen lived, and so on.

While it is a very good system, it depends on the observations and decisions of the classifiers. As time passes, new information might become available that confirms the original classification. That new information might also refute the earlier classification, and a new one is needed. These classifications can also depend on what features the scientist might have considered more important than others. The result of all this is that the taxonomic classification of an organism might change over time. There are even cases where different scientists have wanted to assign different classifications to the same organism, based on seeing things differently. This is certainly the case for the Megalodon.

Early on, various scientists believed that the Megalodon was an ancestor of *Carcharodon carcharias* (genus and species), the modern great white shark. This was largely based on the size and shape of their teeth: large, triangular, and serrated. So, the Megalodon was classified as *Carcharodon Megalodon*. Eventually it was recognized that the serrations on the teeth of these two sharks were very different. The different sizes of the teeth was also very significant. Great white shark teeth have very coarse serrations, while the teeth of Megalodons have very fine serrations. Megalodon teeth also have a very pronounced chevron shaped bourlette between the crown and the root. This bourlette does not exist on great white teeth. So, most scientists agreed that differences between the Megalodon and the great white were too great and that it could not be the ancestor of the great white. Thus, it needed its own genus. The Megalodon was then assigned the new genus of Carcharocles.

It was further determined that the great white and the mako sharks were very closely related and took a different evolutionary path than the Megalodon. Mako sharks belong to the genus Isurus. The *Isurus hastalis*, also known as the broad tooth mako, evolved around 30 mya. Its teeth were wide and up to 3.5 inches (8.9 cm) long and were not serrated. Over time some of those makos started hunting marine mammals. To be more efficient at removing mammal flesh, their teeth started to become partially serrated. These are called transitional teeth because they were transitioning from unserrated to serrated. This transition took place between eight to five mya and was a significant evolutionary event. Those sharks with the partially serrated teeth have been designated *Carcharodon hubbelli* in honor of Gordon Hubbel who originally studied and described them. These same teeth are also sometimes called *Isurus hubbelli* because they were the transition between the Isurus and Carcharodon genus.

As time passed, the transitional teeth became completely serrated with coarse serrations. This led to the identification of an entirely new species, *Carcharodon carcharias*, the great white shark. Fossilized broad tooth mako and great white teeth are very similar in size and shape, with the major difference being that the great whites are serrated and the makos are not. Figure 16 shows the evolutionary sequence from *Isurus hastalis*, to *Carcharodon hubbelli*, to *Carcharodon carcharias*. The serrations on the *Carcharodon hubbelli* are very fine and hard to see in Figure 16. However, if you rub a fingernail along the side of the blade, you will feel them.

The original *Isurus hastalis* did not survive the transition and one mya they became extinct. Relatives of the *Isurus hastalis* did survive and are known today as the short fin and long fin makos. They have unserrated teeth but otherwise look very similar to the modern great white shark.

The fact that the great white is not a descendent of the Megalodon is fine, but one thing is missing. Where did the Megalodon come from? Who were the Megalodon's ancestors? We start down that path by going back 200 mya to the evolution of modern sharks. By that time sharks had evolved and distinguished themselves into several taxonomical orders. One of those orders was Lamniformes which are generally referred to as mackerel sharks. Mackerel sharks are larger predatory sharks with solidly built bodies and pointed snouts. They are also identified by having five gill slits, a mouth that extends behind the eyes, two dorsal fins, and an anal fin. They also have eyes without nictitating membranes. Nictitating membranes are a translucent eyelid that can be moved over the eye to help protect it.

Figure 16. Evolution of the great white shark.

The oldest member of the Lamniformes order, discovered so far, is the *Palaeocarcharias stromeri*. They lived around 165 mya and several sets of their remains have been found in the vicinity of Germany and France.

They were only about three feet (.9 meters) long with a flat body structure, which is not very Lamniformes like. However, their teeth were definitely those of the Lamniformes order as discussed in Chapter 3. There may have been other Lamniformes back then, however they have not been found and identified yet. So, we cannot say that *Palaeocarcharias stromeri* was a direct ancestor of the Megalodon. But, we can say that the Lamniformes definitely were. So, if *Palaeocarcharias stromeri* is not the actual ancestor, it was definitely closely related to them. Figure 17 is an example of one of the *Palaeocarcharias stromeri*.

Figure 17. *Palaeocarcharias stromeri.*

The order Lamniformes has several family members and one of them is the Lamnidae family. The Megalodon belongs to the Lamnidae family. The makos and great whites that we just talked about are also members of the Lamnidae family.

Another Lamnidae, the *Cretalamna appendiculata* evolved and is generally, considered a direct ancestor of the Megalodon. It lived during the Cretaceous, from 103 to 46 mya. By the way, the "creta" in Cretalamna refers to the Cretaceous and all of its chalk deposits. Unfortunately, the sharks that connect *Paleocarcharias stromeri* to *Cretalamna appendiculata* are not well defined. There are too many gaps and questions to create a shark-to-shark lineage. So, it is best to leave it at that until more information is known.

Fortunately, that situation improved greatly with the *Cretalamna appendiculata*. The *Cretalamna appendiculata* was around 9 to 9.8 feet (2.7 to 3 meters) long so it was a medium sized shark. Its teeth were up to one inch (2.5 cm) long and had a wide blade with two cusps. *Cretalamna appendiculata* fossils have been found in Europe, Morocco, West Africa, and North America, to include the east coast and the Midwest, which was covered by the Western Interior Sea during the late Cretaceous. Figure 18 shows a *Cretalamna appendiculata* tooth. This tooth was found in Kazakhstan.

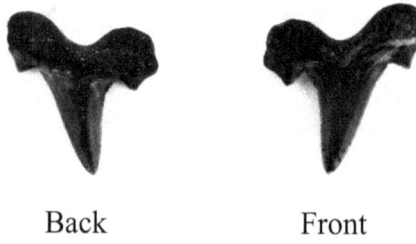

Back   Front

Figure 18. *Cretalamna appendiculata* tooth.

The *Cretalamna appendiculata* evolved into the *Otodus obliquus*, the next direct ancestor of the Megalodon. The *Otodus obliquus* lived from 60 to 45 mya and its fossils have been found worldwide. The Megalodon's ancestors also continued to grow, with the *Otodus obliquus* reaching up to 30 feet (9.1 meters) in length. The largest of its vertebrae centrum ever found was over five inches (12.7 cm) in diameter. Its teeth also continued to grow to over three inches (7.6 cm) long, with the largest ever found being just over four inches (10.1 cm). The teeth look like larger versions of the *Cretalamna appendiculata* with a wider crown and two prominent cusps. Figure 19 shows an *Otodus obliquus* tooth.

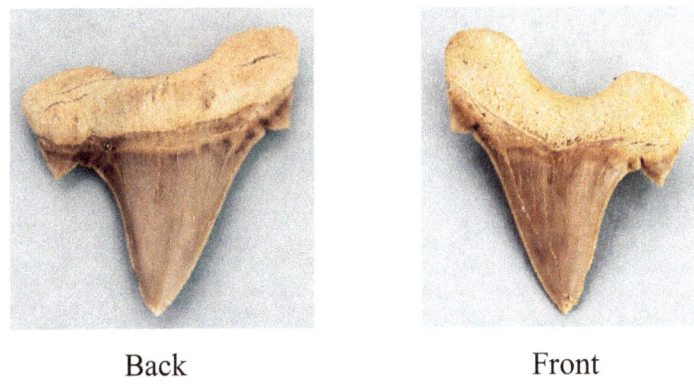

Back   Front

Figure 19. *Otodus obliquus* tooth.

The *Otodus obliquus* evolved into the *Otodus aksuaticus*. These sharks lived from 55 to 49 mya. They were almost identical in size and shape to the *Otodus obliquus*. But, there was one significant difference.

The Otodus teeth were all unserrated until the *Otodus aksuaticus,* which were partially serrated. Obviously, some notable evolution was starting to occur in the Otodus line. The development of serrations in shark teeth is an evolutionary response to eating larger prey, such as marine mammals. Figure 20 shows an example of an *Otodus aksuaticus* tooth. The serrations on the crown of this photograph are a bit hard to see but you can readily see serrations on the cusps.

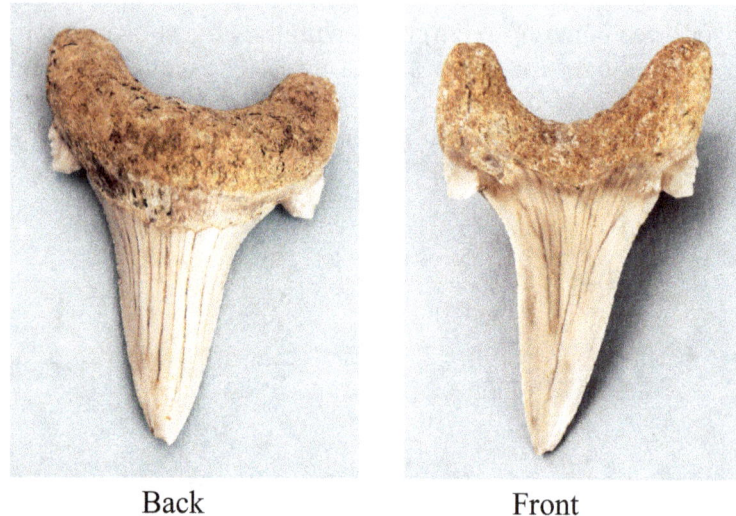

Back   Front

Figure 20. *Otodus aksuaticus* tooth.

    The development of those serrations has led to a very recent discussion in the taxonomy of the Megalodon ancestors. Some scientists believe that the serrations on the *Otodus aksuaticus* were a normal evolutionary development of the genus Otodus and that all of the evolutions that followed should be considered to be Otodus. This includes all the way to the Megalodon. Other scientists believe that the serrations on the *Otodus aksuaticus* were more than a normal evolution, and that it was a transitional shark to a new genus: Carcharocles. So, you can easily see how different interpretations of a characteristic can lead to a potential change in the taxonomy of an organism. This is why there are two emerging points of view in the Megalodon world. Should it be called the *Otodus Megalodon* or the *Carcharocles Megalodon*? These kinds of diversions are not uncommon in taxonomy.

    The discussions about *Carcharocles Megalodon* or

Otodus Megalodon (and their ancestors) are far from over. Personally, I think that the Carcharocles argument is correct. A somewhat similar situation is what we discussed earlier with the evolution of the great white sharks. In that case they were named as the new species *Carcharodon carcharias* after the transition. I can see the pros and cons for both points of view. But, having mentioned that, for now I have to choose which one to use, Otodus or Carcharocles. So, for purposes of this book, I will use the genus Carcharocles.

This means that the *Otodus aksuaticus* evolved into the *Carcharocles auriculatus*. The *Carcharocles auriculatus* existed 38 to 25 mya. It could be found worldwide in temperate waters and grew to 30 feet (9.1 meters) long. Their teeth were up to 4.5 inches (11.4 cm) long, narrow, fully serrated, and had cusps. Figure 21 shows a *Carcharocles auriculatus* tooth.

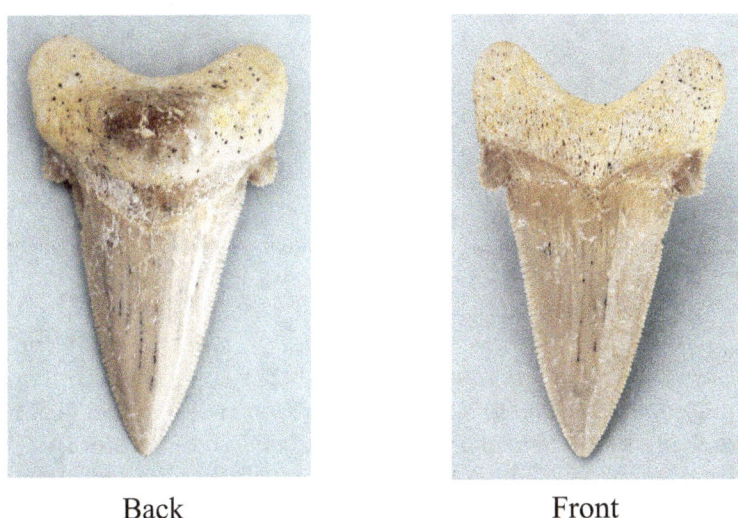

Back   Front

Figure 21. *Carcharocles auriculatus* tooth.

The *Carcharocles auriculatus* evolved into the *Carcharocles angustidens*. The *Carcharocles angustidens* lived from 33 to 22 mya, and, like its predecessors, it could be found worldwide in temperate waters. It grew to 30 to 35 feet (9.1 to 9.5 meters) long with teeth that reached up to 4.5 inches (11.4 cm) long. The teeth of the *Carcharocles angustidens* were wider but had smaller cusps than *Carcharocles angustidens*. Figure 22 shows a *Carcharocles angustidens* tooth.

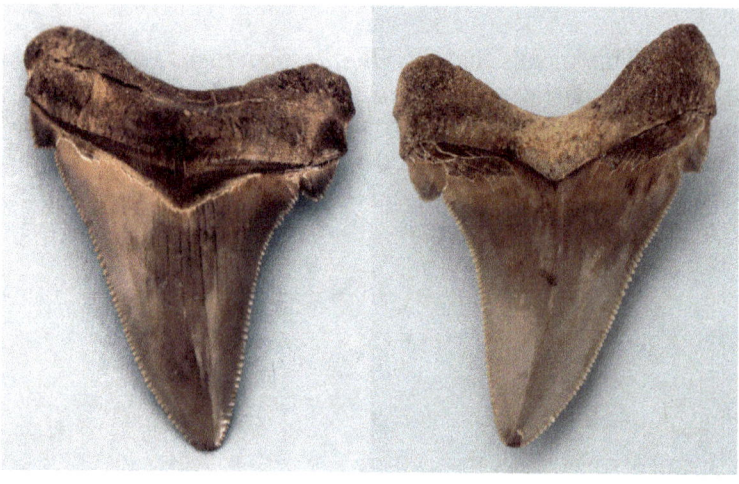

Back   Front

Figure 22. *Carcharocles angustidens* tooth.

The *Carcharocles angustidens* evolved into the *Carcharocles chubutensis*. These sharks lived 28 to 5 mya in temperate waters worldwide. They grew to 50 feet (15.2 meters) long. *Carcharocles chubutensis* teeth were up to five inches (12.7 cm) long, wide, and fully serrated. The cusps of its predecessors had almost vanished with only barely noticeable vestigial cusps left. Figure 23 shows a *Carcharocles chubutensis* tooth.

Vestigial cusps

Back

Figure 23. *Carcharocles chubutensis* tooth. (Front on the following page)

MEGALODON!

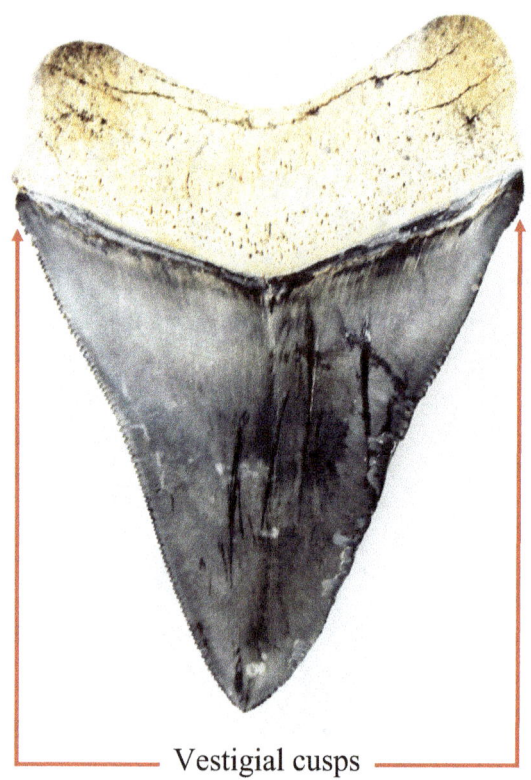

Vestigial cusps

Front

Figure 23. *Carcharocles chubutensis* tooth.

    The *Carcharocles chubutensis* evolved into the *Carcharocles Megalodon*. The Megalodon lived in temperate waters worldwide from 23 to 2.6 mya. It reached up to 60 feet (18.3 meters) long and its teeth were wide and long and reached over six inches (15.2 cm) in length. A few Megalodon teeth have been found just over seven inches (17.8 cm) long. All Megalodon teeth have very fine serrations and no cusps. Figure 24 shows a *Carcharocles Megalodon* tooth.

Back

Figure 24. *Carcharocles Megalodon* tooth.
(Front on the following page)

Front

Figure 24. *Carcharocles Megalodon* tooth.

    The above descriptions spoke of how the *Carcharocles auriculatus* evolved into the *Carcharocles angustidens* and then into the *Carcharocles chubutensis* and then into the *Carcharocles Megalodon*. In fact, there was not a smooth transition of one into the other. They actually evolved alongside each other with the predecessor finally dying off leaving the newer version to carry on, and, for the process of evolution to continue until we reached the *Carcharocles Megalodon*. The dates associated with when the Megalodon's ancestors were alive are also a bit uncertain. Most publications only designate the geologic epoch when they existed, possibly including early, mid, late, such as the middle Eocene. You will probably see other information online or in books that might differ somewhat from what I have written. All that said, when the last Megalodon died, that was the end of the line for all those amazing sharks.

This is a lot of information to consider. So, to help put it all into perspective, Table 3 summarizes the known ancestors of the Megalodon. Figure 25 captures the whole evolution including the teeth, names, and how many years ago they all lived.

Shark	**MYA**
First primitive sharks	416
Modern sharks evolved	200
Paleocarcharias stromeri (may or may not be a direct ancestor)	165
Cretalamna appendiculata	103-46
Otodus obliquus	60-45
Otodus aksuaticus	55-38
Carcharocles auriculatus	38-25
Carcharocles angustidens	33-22
Carcharocles chubutensis	28-5
Carcharocles Megalodon	23-2.6

Table 3. Ancestors of the *Carcharocles Megalodon*.

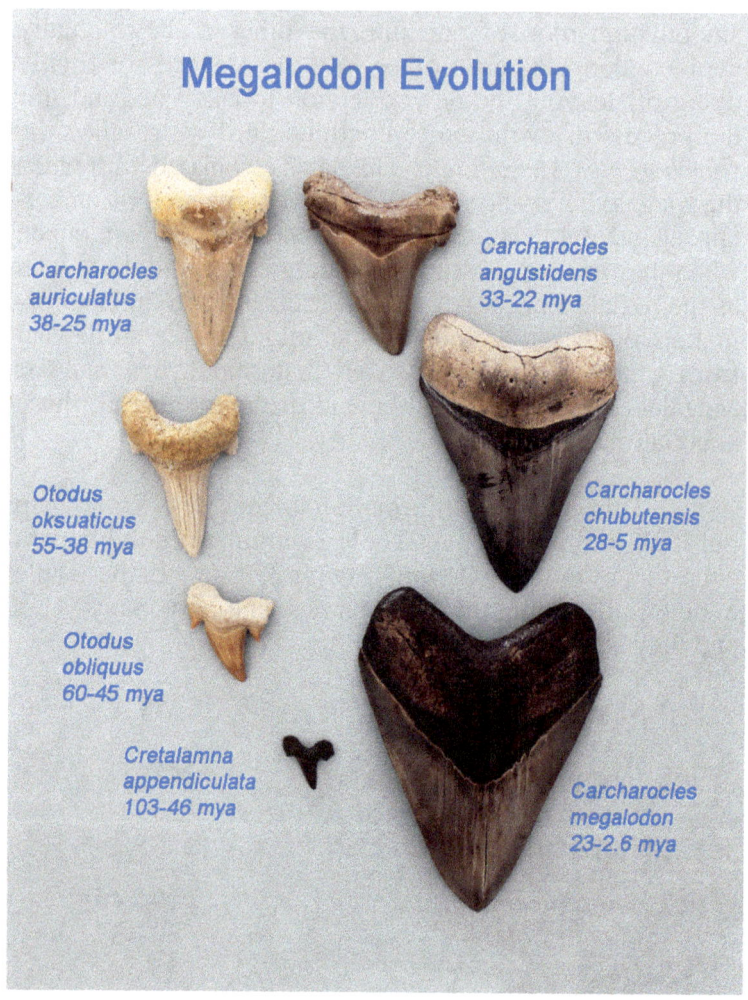

Figure 25. Evolution of the *Carcharocles Megalodon*.

The Megalodon became extinct around 2.6 mya. There are many theories as to why this happened. There are two that seem to withstand the test of time. One is that the Megalodon had to eat very large animals, generally whales, to remain alive. Whales could tolerate much colder water than Megalodons, creating a negative impact on the Megalodon's diet when the whales performed their annual migration to colder waters. The second theory, which probably had a much greater impact, was that sharks in general were abundant when the Megalodons were alive. Many of these sharks were large, with the great white being the largest of them all. These large sharks were much more maneuverable and could out compete the Megalodons for food. In the end, the Megalodons reached a biological dead end and died off.

Figure 26 is a good comparison of a 5.75 inch (14.6 cm) Megalodon tooth and a 1.5 inch (3.8 cm) great white tooth. The Megalodon is considerably larger with very fine serrations and a bourlette. The great white is much smaller with more coarse serrations and no bourlette. These differences will help you determine if the tooth you found is a large great white or a small Megalodon.

Figure 26. Comparison of a Megalodon and a great white tooth.

Chapter 6

The Megalodon Body

As mentioned in Chapter 5, Megalodons are a member of the Lamniformes order, which are generally referred to as mackerel sharks. Mackerel sharks are larger predatory sharks with solidly built bodies and pointed snouts. They are also identified by having five gill slits, a mouth that extends behind the eyes, two dorsal fins, and an anal fin. Great white sharks and Megalodons belong to the Lamniformes order. So, we can reasonably assume that the Megalodon looked more or less like a 60-foot-long (18.3 meters) version of a great white shark. Figure 27 shows the presumed general shape and fin arrangement of the Megalodon.

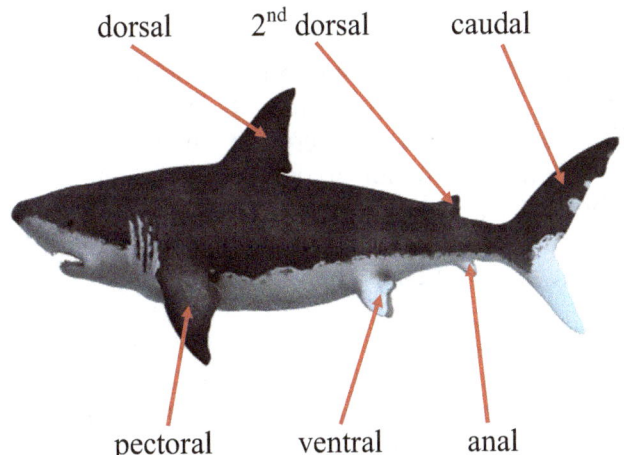

Figure 27. General shape and fins of the Megalodon.

Fins

Extrapolating from a great white shark's fins, it is possible to figure out a range of lengths of Megalodon fins. The iconic dorsal fin on the Megalodon's back would be around seven feet (2.1 meters) tall. This fin helps provide stability. The caudal (tail) fin would be around 15 to 20 feet (4.6 to 6.1 meters) tall from top to bottom and provides propulsion. The vertebrae centra run through the top part of the caudal fin, so it is always larger than the bottom part. The pectoral fins would be about 13 feet (4 meters) long and provide maneuverability up, down, right, and left. The second dorsal fin provides additional stability and maneuverability to the back end of the shark. The anal fins add additional stability. The ventral fin helps the shark turn and roll. The dorsal fins and anal fins would probably be one to two feet (.3 to .6 meters) long. The ventral fins would probably be about three feet (.9 meters) long.

Cartilage

Cartilage is a strong and flexible material that is commonly found in our bodies. Good examples of our cartilage are our noses and sternum, which is the front part of our rib cage where our rib bones come together. Take a deep breath and notice your rib cage expanding. The cartilage in our sternum is largely responsible for our rib cage being able to expand and contract easily.

The cartilage in a shark's body does not exactly match the bones in a human body. Figure 28 shows what a Megalodon's cartilage skeleton would probably have looked like. This is a drawing of a great white shark skeleton made in 1887. For being over 130 years old, it is remarkably informative. The cartilage structures behind the head and in front of the pectoral fins look similar to ribs, but they are not ribs. They actually provide support to the gills. The long angled piece just to the left of them provides support for the pectoral fins. The other fins have internal cartilage structures. Everything is supported by the long vertebral column. The vertebrae are also made of cartilage and are big, thick, round disks called centra. Figure 29 shows a section of fossilized centra from an *Otodus Obliquus*. The largest of them is five inches (12.7 cm) in diameter. Megalodon centra are almost identical in appearance and up to nine inches (22.9 cm) in diameter. The small, white disk in the lower left is a modern tiger shark vertebrae for comparison.

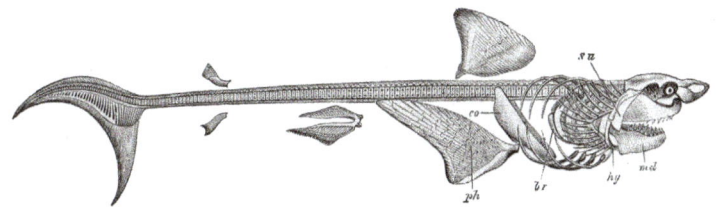

Figure 28. Megalodon's cartilage skeleton.

Figure 29. Fossilized *Otodus Obliquus* centra.

Fossilized Megalodon remains

Very few sets of partial remains of Megalodons have been found. In 1926, the most complete set of Megalodon vertebra centra found so far was discovered in the Antwerp Basin of Belgium. There were 150 centra notated ranging from 2.2 to 6 inches (5.6 to 15.2 cm) in diameter. It is believed that the Megalodon actually had more centra, but they were not found at the site. With 150 centra this Megalodon would have been around 52 feet (15.8 meters) long. It would have been even longer if there actually were missing centra.

In 1983, one broken Megalodon tooth and 20 centra were found in the Danish Upper Miocene Gram Formation in Denmark. The broken tooth was 5.9 inches (15 cm) long and, if a missing piece had been present, the tooth would have been 6.3 inches (16 cm) long. The centra ranged from 3.9 to 9.0 inches (9.9 to 22.9 cm) in diameter.

In 2001, a well-preserved *Carcharocles angustidens* was found in New Zealand's South Island. This set of remains had 165 teeth and 32 vertebral centra. It is estimated that this shark was 31 feet (9.5 meters) long and the biggest teeth were 3.9 inches (9.9 cm) long.

Skull and Jaws

Starting at the head, we see that the skull and jaws look very different from our human head and jaws. Part of the reason for this is that sharks are normally oriented horizontally, in the water, and not vertically and on land like humans. Our skull sits on top of a spinal column. In the shark the skull is a horizontal extension of their spinal column. A shark's head also has to be streamlined to swim efficiently in the ocean.

Strong and flexible connective tissue connects the upper jaw with the lower jaw and both jaws to the skull. That connective tissue allows the jaws to open wide and extend forward to achieve a larger bite. A full-grown Megalodon would have a mouth opening of about six feet by six feet (1.8 by 1.8 meters). Figure 30 shows a drawing of a Megalodon skull and jaws based on a great white skull. The rostrum in this figure is just the structure behind the pointy end of the shark's head. The brain and the Megalodon's organs will be discussed in Chapter 7, Megalodon Sensory And Internal Organs.

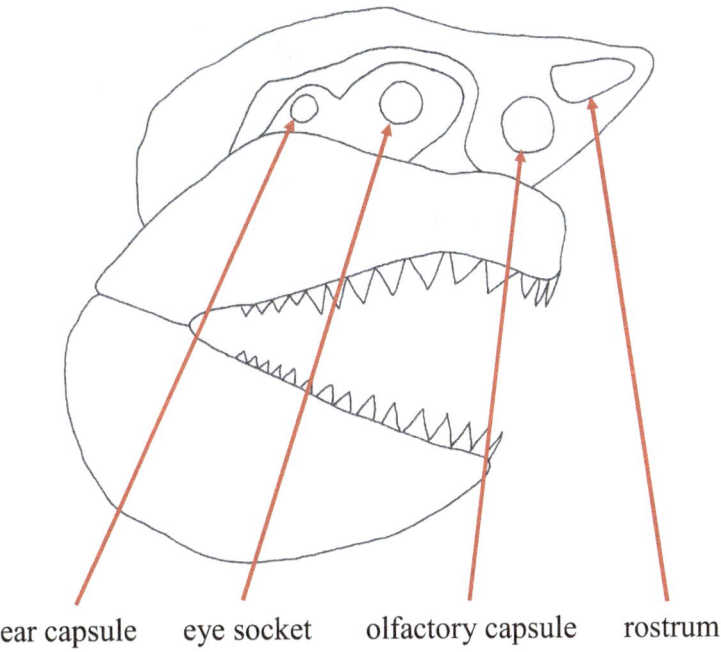

ear capsule   eye socket   olfactory capsule   rostrum

Figure 30. Megalodon skull and jaws.

Weight
    A quick analysis can give us a rough estimate of how much a Megalodon weighed. The larger that an animals is, the more they weigh compared to extrapolated versions of smaller animals of the same or similar species. That is because we are not just comparing length. We also have to factor in the increases in height and width. So, trying to do a one-dimensional extrapolation from a great white would not be accurate. But, there are two examples we can work with. Whale sharks can reach up to 60 feet

MEGALODON!

(18.3 meters) long and at that length they weigh about 94,800 pounds (43,000 kg). That is an overall average of about 1500 pounds per foot (2235 kg per meter). Humpback whales can reach up to 56 feet (17 meters) and 88,000 pounds (39,916 kg) for an overall average of about 1570 pounds per foot (2339 kg per meter). Using these comparisons, a 60-foot (18.3 meter) Megalodon would weigh approximately 90,000 to 94,200 pounds (40,823 to 42728 kg). By contrast, a great white shark can reach 20 feet and weigh 4,200 pounds (1905 kg) for an overall average of 210 pounds per foot (313 kg per meter). As you can see, that is considerably less than the pounds per foot (kg per meter) for much larger animals, which is why we did not extrapolate from the great white in this instance.

Chapter 7

Megalodon Sensory And Internal Organs

This chapter is an overview of the Megalodon's sensory and internal organs. No one has ever seen these organs, so we don't really know what they look like or how big they are. But, we know about the Lamniformes and particularly the great white shark. So, the following information is for a fully grown 60 foot (18.3 meter) long Megalodon, based on comparison and extrapolation with a fully-grown 20 foot (6.1 meter) long great white shark. It might be a bit of a stretch to make this kind of comparison, but, it will be in the ball park and will definitely help you get a better understanding of what the Megalodon was probably like.

As we saw in the previous chapter, sharks do not have a rib cage like we do to protect our heart, lungs, and other internal organs. Shark organs and some connective tissue are all neatly tucked inside their muscles and a very tough, streamlined skin. They do, however, have some "rib like" cartilage that provides structure to their gills and support the pectoral fins.

Brain

Since sharks have long fusiform (tapered toward both ends) bodies, their skull, brain, and all of their organs have to fit in this streamlined structure. We saw the skull in the previous chapter, so now we will look at the brain. Based on a great white shark, a full-sized Megalodon would have a brain that is approximately 28 inches (71.1 cm) long and 5 inches (12.7 cm) wide. Figure 31 is a drawing of such a brain with the associated olfactory bulbs, eyes, otic (ear) capsules, and their nerve connections. The nerves that connect to the electroreception system (lateral line and the Ampullae of Lorenzini) are also shown. All the other nerves that connect to the rest of the body are not detailed.

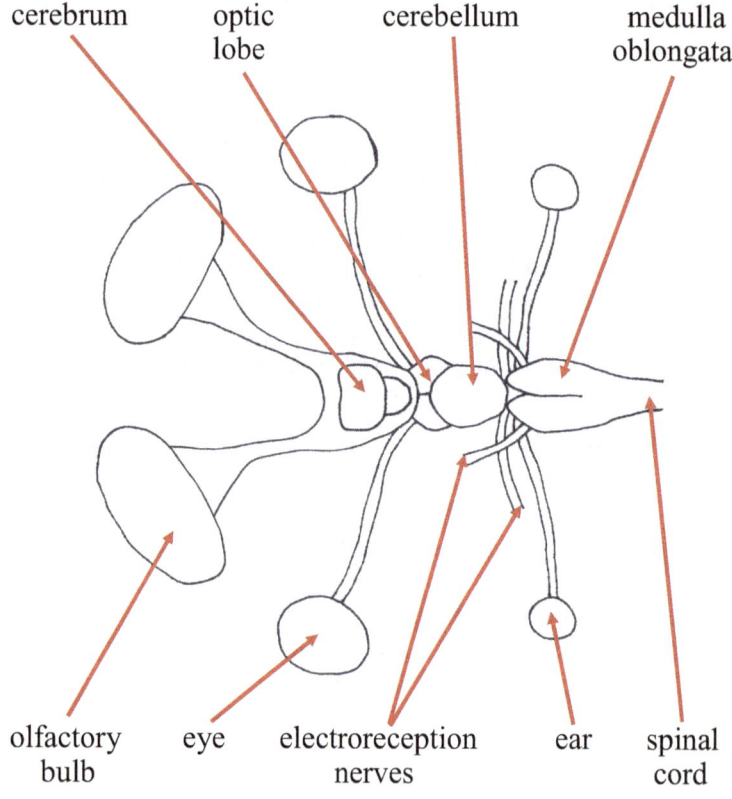

Figure 31. Drawing of Megalodon brain.

Smell
    Extending out from the front of the brain are long nerves leading to the large olfactory bulbs. Between a shark's eyes and tip of their snout you can see two small openings in their outer skin. These openings allow water to flow past the olfactory bulbs and let the shark smell what is in the

water. By smell, we actually mean they can detect chemicals and molecules in the water, such as blood. Great white sharks have the largest olfactory bulbs of any living shark, about four inches (10 cm) in diameter. The Megalodon would have had olfactory bulbs around 12 inches (30.5 cm) in diameter.

Sight

Megalodon eyes would have been approximately nine inches (22.9 cm) in diameter. Since Megalodons are part of the Lamniformes order, their eyes would not have had nictitating tissue. Nictitating tissue is translucent tissue that a shark can move over its eye for protection when feeding, or any other time it is needed. Instead, Lamniformes have eyes that the shark can rotate back in their eye socket for protection. You may have seen videos of great white sharks feeding and noticed that their eyes roll back so that you can only see dense, white tissue. This tissue is actually the back side of their eye.

Hearing

A short distance behind each eye is a small hole for hearing. The hole allows sound, pressure waves in the water, to reach their otic capsule. This capsule is a cartilage structure with three semicircular chambers and an otolith (sometimes called an ear stone). The semicircular chambers help the shark maintain balance and the otolith does the hearing. The nerves from the ear connect to the posterior section of the brain. The otic capsule is a little smaller than the eye.

Electroreception system

Sharks have an electroreception sensory system that helps them stay aware of their environment. This system is comprised of the lateral line and the Ampullae of Lorenzini.

The lateral line runs the length of the shark's body from the side of the head to the tail and is lined with cells that allow them to detect pressure changes in the water. In addition to the general flow of water, the lateral line can detect when another animal is swimming nearby. So, this system helps keep sharks alert to what is going on around them, especially things that are out of their eyesight or in the darkness.

The Ampullae of Lorenzini are electroreceptor organs that allow the shark to detect changes in the local electrical field. On the snout and face of the shark are large numbers of gel-filled pores. The nerves in these pores detect changes in the local electrical field caused by living creatures nearby. Hammerhead sharks use these receptors to find prey, such as stingrays, buried under the sand. So, the Ampullae of Lorenzini help sharks maintain awareness of their surroundings. Figure 32 shows the Ampullae of Lorenzini on the snout of a great white shark. They look kind of like freckles. In this figure, you can also see the small, dark, nostrils opening to the olfactory bulbs. The left eye is visible on the side of the shark's head. In this figure the Ampullae of Lorenzini are best seen in the darker skin area above the nostrils.

Sharks can also detect fat in their prey. Fat is a major source of nourishment and energy for sharks. Whales, seals, walruses and other marine mammals have thick layers of fat and blubber to store energy and provide insulation from cold waters. If a shark is uncertain whether a possible prey item has enough fat to be worth the effort of killing and eating, it may perform an exploratory bite. That means that when the shark bites the potential prey, if it detects enough fat it will continue to kill and eat it. If it does not de-

Megalodon Sensory And Internal Organs

Figure 32. Ampullae of Lorenzini

tect enough fat, it may abandon the potential prey and resume hunting. This explains why many humans suffer a single shark bite and the shark leaves the area even though there is a lot of blood in the water.

Other organs

Moving beyond the skull and brain, we can talk about some of the Megalodon's major organs. The heart is long and thin and sort of "S" shaped. It is located in the lower part of the body just behind the gills. Unlike our four-chambered hearts, this heart has two chambers. A full grown Megalodon would have had a heart about four feet (1.2 meters) long.

The liver is a very important organ to any shark. It contains lots of oil and can provide nutrition to the shark during periods when there is no prey available. It is also used to help control buoyancy. Great white shark livers occupy about one third of their body length. So, a fully-grown Megalodon could have a liver about 20 feet long (6.1 meters).

The stomach of a Megalodon would be about 10 to 15 feet (3 to 4.6 meters) long. It would also be able to stretch in girth. So, a Megalodon would be able to ingest huge amounts of meat.

Once processed in the stomach, the food that was eaten moves on to the intestines. Most animals have a long, tubular intestine that folds back and forth in the abdomen to provide a large surface area to absorb nutrients from their food. Sharks have a different arrangement. Their intestines are a long tube with a spiral absorption structure inside, looking much like a grain auger that a farmer would use. The auger part of the intestine provides the large surface area to absorb nutrients from the food. When the intestinal waste leaves the shark's body, it comes out in a spiral shape. A Megalodon's intestine would be about nine feet (2.7 meters) long.

Reproduction

In addition to all their other fins, male great white sharks have a two-piece fin like structure called claspers. They are the male shark's external sexual organs and they are located on the underside, along the center line, and just behind the ventral fins. It is assumed that male Megalodons also had claspers.

Female great white sharks reach sexual maturity at about 33 years old. They would be quite large by then. The largest number of great white pups recorded is 14. Without additional information we can only assume that the Megalodon had a similar number of pups. Analysis of growth rings on fossilized Megalodon centra, compared with its further growth and age, indicates that Megalodon pups were around 6.6 feet (2 meters) long at birth.

Great white sharks are ovoviviparous, which means that their eggs hatch while still in the mother's uterus. As the hatched pups continue to grow in the uterus, they feed on eggs that their mother has produced.

Before we finish our discussion of the sensory and internal organs, we can take an interesting little detour. Do sharks have tongues? The answer is "yes," but not like the tongues of most other animals. Most tongues are soft and flexible so that they can move food around during chewing and swallowing. For sharks, the tongue is tougher and not very flexible. It lies on the bottom of the mouth with muscles on each side that can lift its front. Take another look at Figure 32 which showed Ampullae of Lorenzini. Laying on the bottom of the open mouth is a large, flat, pink tongue with several muscles on the side. The tongue acts much like a large scoop that helps send down the throat whatever it has bitten off. As discussed earlier in this book, the back "display" sides of a shark teeth are rounded and attractive looking. When the shark bites its prey, that rounded surface immediately starts moving the flesh away from the prey and deeper into the mouth. Then the tongue takes over to complete the process. A Megalodon's tongue would have been around four feet (1.2 meters) long.

Chapter 8

Where Megalodon Teeth Have Been Found

If you are reading this book, you are probably keenly interested in finding teeth from the Megalodon and its ancestors. They are not easy to find, but with some hard work and study, it is possible. I have found all of my Megalodon teeth while SCUBA diving in the Gulf Of Mexico off Venice, Florida. Venice is a hot spot, but there are many other hot spots in the world.

Later in this chapter is a quick overview of locations around the world where teeth from the Megalodon and its ancestors have been found. It is not meant to be a specific guide of exactly where to go to find Megalodon teeth, but I will include some detailed information when it is helpful. Unfortunately, many locations are now off-limits for collecting, especially overseas. It is also true that some sites can change their rules at any time. So, if you are interested in actually visiting any of these areas, make sure that you do your research in advance to know if it is even possible to look for teeth there. If you want more information, there are many resources that should be able to help. A good place to start is with the local fossil clubs. If hunting for Megalodon teeth would involve diving, contact the local SCUBA shops and dive clubs. Many fossil hunting SCUBA divers are very knowledgeable about fossils and shark teeth. Other places to look are the local and national government web sites. There are several books that talk about hunting Megalodon teeth. You can also find lots of information online, but always beware that not everyone out there is a reliable source of information. So, do your due diligence and don't make travel plans based on possibly erroneous information.

Megalodon teeth are rare and hard to find. Finding one in the sand on the beach is extremely rare, but does happen. Mostly they are found in the ocean, rivers, and creeks. They are also found in surface mines, quarries, and construction sites. So, be prepared for some measure of physical activity. If it was easy, everyone would have one. When you do find one it will be extra special because you studied and put in the work to make it happen. You might want to read my book *Hunting Fossil Shark Teeth In Venice, Florida*. Even though it is focused on Venice, Florida, most of the information and tips apply to almost any site, on land or under water, that you might be interested in exploring.

Some locations require a permit to search on public lands. For example, Florida requires a state permit to search for any fossil, except shark teeth, on public lands. Other states are more restrictive. Some foreign countries are very restrictive. Every country, state, and local area is different, so check this out before you go.

Be aware of and respect private property. For example, one of my favorite places to hunt was Calvert Cliffs along the west side of the Chesapeake Bay in Maryland. In the "old days," we could drive right up to the beach in any of a number of small towns and head right to the water. Since then, all of those communities have instituted locals-only regulations. This means that unless you live there, or are a guest of a resident, you are not allowed access to their beach. If you had not hunted there for many years you would be in for a big surprise when you returned, thinking that nothing had changed. One way around this is to get to the area by boat and dive there. Also, don't dig in the banks and cliffs if you do have permission to visit the site. That causes erosion, which is bad for nature and the property owners who have houses on top of those cliffs. While on the subject, don't dig in the banks of any river or creek because of the erosion it causes.

If you visit a surface mine or construction site, be sure to arrange permission from the owner or foreman. Many of these places are posted "No Trespassing" under penalty of a felony. Most rivers and creeks are OK as long as you stay in the water and away from private property.

Now lets look at why certain areas are good and others are not. We start by reflecting on information that was covered in Chapters 4 and 5. For convenience, Table 3 from Chapter 5 is repeated below.

Shark	MYA
First primitive sharks	416
Modern sharks evolved	200
Paleocarcharias stromeri	165
(may or may not be a direct ancestor)	
Cretalamna appendiculata	103-46
Otodus obliquus	60-45
Otodus aksuaticus	55-38
Carcharocles auriculatus	38-25
Carcharocles angustidens	33-22
Carcharocles chubutensis	28-5
Carcharocles Megalodon	23-2.6

Table 3. Ancestors of the *Carcharocles Megalodon*.

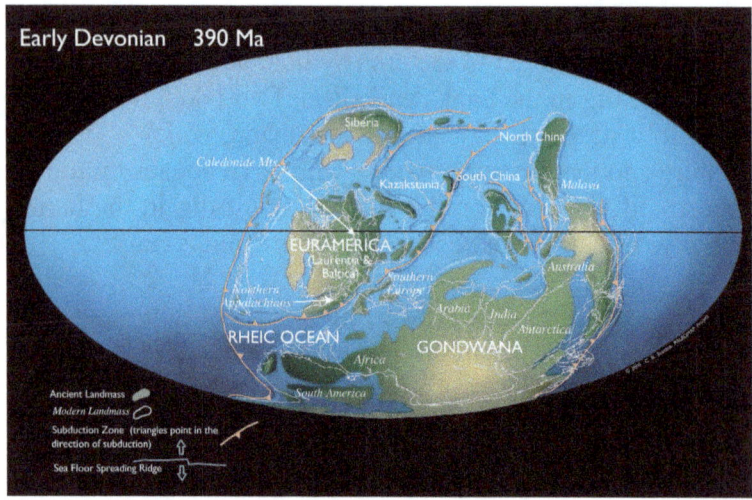

Figure 33. Earth 390 mya.

    The first primitive sharks emerged around 416 mya. Figure 33 is a paleogeographic map that shows what the Earth looked like 390 mya. As you can see, it was very different from the world we know today. In this map, and the maps that are presented below, you will see the effects of tectonic plate movement and how the land masses and water levels have continued to change over time.

    Modern sharks emerged around 200 mya. *Palaeocarcharias stromeri* evolved around 165 mya in the vicinity of Germany and France. So, let's take a look at what the Earth looked like back then. Figure 34 is a map of the Earth 152 mya. As you can see, most of Europe, to include France and Germany, were covered by the oceans. Based on where their fossils have been found, we know that *Carcharocles Megalodon* and its ancestors were largely coastal sharks in temperate areas of the world. So, when you look at these maps, look for the coastal areas at those

Where Megalodon Teeth Have Been Found

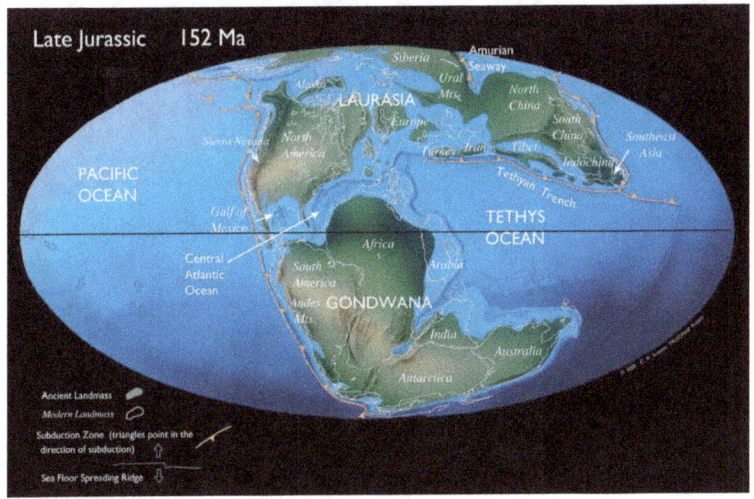

Figure 34. Earth 152 mya.

times. Those are the locations where the Megalodon and its ancestors roamed and their fossils can be found.

About 103 to 46 mya the *Cretalamna appendiculata* lived, and its fossils have been found, in Europe, Kazakhstan, Morocco, West Africa, and North America, to include the east coast and the Western Interior Sea in the Midwest. Figure 35 is a very interesting map that shows what the Earth looked like 94 mya with the Western Interior Sea, which divided North America in two. It is easy to see where the *Cretalamna appendiculata* lived and why its fossils are found in those various locations.

Next came the *Otodus obliquus* and *Otodus aksuaticus*, which lived 60 to 45 mya and 55 to 38 mya respectively. Figure 36 shows what the Earth looked like 50 mya. More and more land has emerged but there were still significant ocean covering what would be the modern coastal areas.

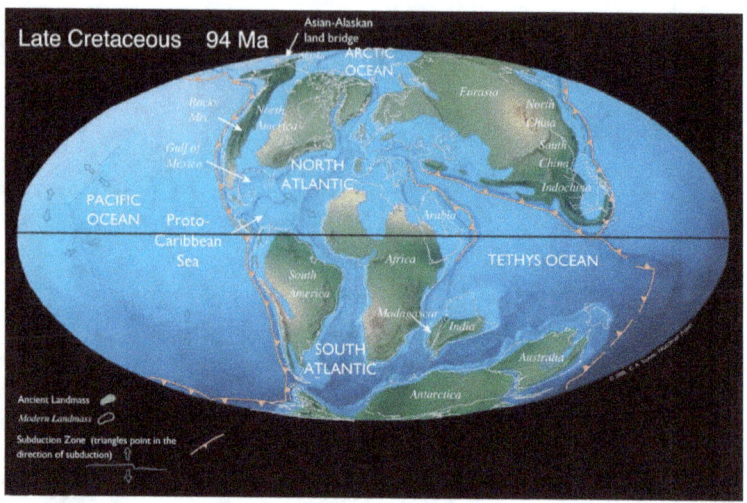

Figure 35. Earth 94 mya.

The *Carcharocles auriculatus* lived from 38 to 25 mya and the *Carcharocles angustidens* lived from 33 to 22 mya, which was the late Eocene to the early Miocene. Figure 37 shows the Earth 14 mya during the middle Miocene. The *Carcharocles auriculatus* and the *Carcharocles angustidens* lived in the time frame between Figure 36 and Figure 37. The main difference between these two figures is that more and more coastal land was emerging. So, generally if you are looking for teeth from these two sharks you generally need to be looking further inland than the modern coasts.

The *Carcharocles chubutensis* lived from 28 to 5 mya and the *Carcharocles Megalodon* lived from 23 to 2.6 mya. Figure 37 shows the Earth 14 mya, right in the middle of when these two sharks were alive.

Where Megalodon Teeth Have Been Found

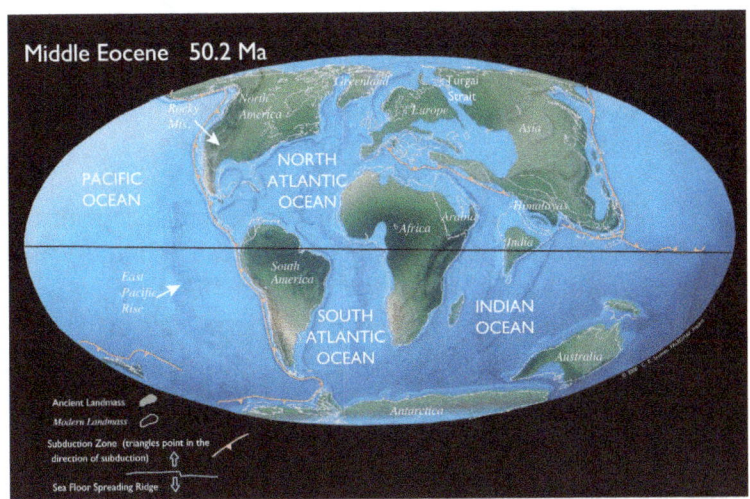

Figure 36. Earth 50 mya.

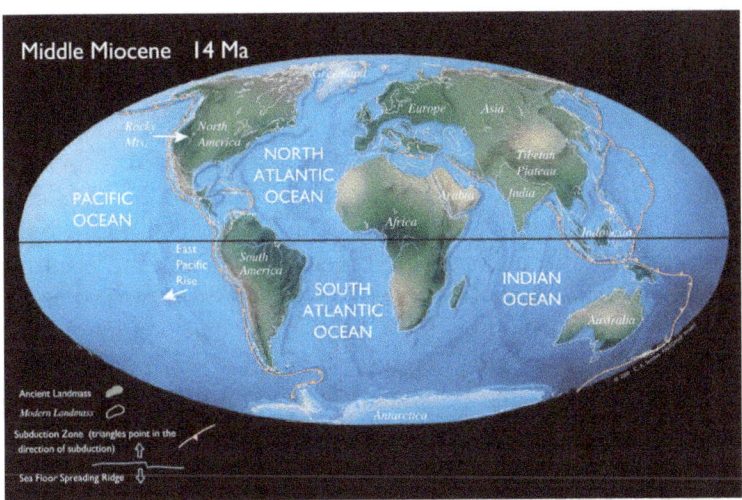

Figure 37. Earth 14 mya.

So, when looking for Megalodon teeth, it is best to focus on areas that were coastal during the Miocene. A very good example of this is the fact that Megalodon teeth can be found all the way from Florida to New Jersey.

The following is a list of sites where the teeth of the Megalodon and its ancestors have been found. With the background information presented above, this list will be more meaningful. Of course, this is not an extensive list of places, which would be too much to document. So, this is a list of the major locations. Hopefully, I did not miss too many. Some of these sites are off-limits to collecting. But, the main purpose of this list is to show where teeth have been found to help further your understanding of where the Megalodon and its ancestors roamed.

United States
California - Bakersfield and Shark Tooth Hill.

Florida - Megalodon teeth can be found in most of the northern half of the state. The southern end of this area is where the Peace River empties into the Gulf of Mexico. The coastal areas on each side of the central Florida ridge are good, especially in the rivers that drain the central highlands. Venice is especially good. Look in the Peace River all the way from the Gulf to the head waters above Wauchula. Other areas of interest include Gainesville, Fernandina Beach/Amelia Island, Jacksonville Beach, Saint Mary's River, Santa Fe River, Suwannee River, Ecofina River, the Withlacoochee River and too many other sites to mention.

Georgia - Megalodon teeth can be found in numerous places along Georgia's coast. They can also be found by diving in the rivers and bays in the south east. They have

Where Megalodon Teeth Have Been Found

been found in the Savannah River and also in the spoil islands that were created from dredging that river.

Maryland - The most well-known area is Calvert Cliffs on the west side of the Chesapeake Bay. Another good place to look is on the banks of the lower Potomac River.

New Jersey - Megalodon teeth have been found in Cumberland county, which borders the Delaware Bay. They have also been found in Big Brook and the Shark River Park.

North Carolina - Megalodon teeth have been found on the beaches, but hot spots are the greater Beaufort and Pit County areas. This is the location of the very famous Aurora/Lee creek phosphate mine. They can also be found in the ocean on ledges 27 to 42 miles (43 to 68 kilometers) offshore in 100 feet (30.5 meters) of water. The teeth here washed out of the Pungo River formation.

South Carolina - Megalodon teeth have been found on the beaches, especially Myrtle and Folly Beaches. Hot spots are diving in the Cooper River, Morgan River, and several other local rivers. Summerville in the low country is also very good for inland searching, especially the Chandler Bridge Creek area. Megalodons are also found in the ocean, similar to North Carolina.

Texas - Much of eastern and southern Texas was underwater until the late Miocene. As a result, teeth from the Megalodon and its ancestors, to include *Cretalamna appendiculata,* have been found there. Some Megalodon teeth have been found on the beaches around Galveston. Another popular location is the Post Oak Creek riverbed in Sherman.

Virginia - The lower Potomac River, to include Belvedere Beach, Nomini Cliffs, and Shark Tooth Beach, are popular. Megalodon teeth have also been found in the York River.

Louisiana, Arkansas, Alabama, Florida panhandle - There are very few reports of Megalodon teeth being found in these areas. However, since these areas were underwater during most of the Miocene, and before, it is not out of the question that Megalodon teeth are out there somewhere, perhaps in the Gulf of Mexico.

Caribbean
Caribbean - Some Megalodon teeth have been found in quarry sites.

Cuba - Some Megalodon teeth have been found in quarry sites.

Mexico, Central and South America
Argentina - Some Megalodon teeth have been found in the Punta Medanos area of the Buenos Aires Province.

Chili - One of the largest accumulations of Megalodon teeth in the world is in the Atacama desert. This area is in the northern end of Chili.

Mexico - Some Megalodon teeth have been found in Baja. Some have also been found in water-filled cenotes (caverns) in the Yucatan.

Panama - Numerous Megalodon teeth have been found here. This area was once open and separated North and South America, allowing the Caribbean Sea and the Pacific Ocean to mix. The Isthmus of Panama formed in the middle-Miocene, closing off this connection.

Peru - The Ocucaje desert is another area with large accumulations of Megalodon teeth and many of them are very large. One of the contenders for the largest Megalodon tooth ever found is 7.48 inches and was found here.

Africa
Madagascar - Some Megalodon teeth have been found here.

Morocco - Many Megalodon teeth have been found here. Some of its ancestors including the *Otodus obliquus* are commonly found in the phosphate mine around the Khouribga area.

Asia/Pacific
Japan - A few Megalodon teeth have been found here.

Indonesia - Large numbers of Megalodon teeth have been found in West Java. They are found in hill sides and stream beds.

New Caledonia - This is a small island between Fiji and Australia. Teeth have been found by dredging in the ocean in that area.

Australia - Megalodon teeth have been found here, especially at the Cape Range National Park on the western coast.

New Zealand - Angustidens remains have been found on the South Island.

Europe
Belgium - Megalodon remains have been found in the Ant-

werp Basin. They have also been found in numerous quarries. The border between Belgium and the Netherlands has also produced many Megalodon teeth.

Denmark - Megalodon remains and teeth have been found here.

England - Megalodon teeth are sometimes found on the beach on the Suffolk coast and in Essex.

France - Megalodon teeth have been found in Savigny and Touraine.

Kazakhstan - Some Megalodon teeth have been found here. Additionally, some of its ancestors such as the *Otodus obliquus* and *Otodus Aksuaticus* can also be found.

Malta - Some Megalodon teeth have been found there.

Netherlands - Megalodon teeth have been found on the beaches of Zeeuws-Vlaanderen, and the border with Belgium.

Other places in Europe - There are reports of Megalodon teeth having been found in Holland, Italy, Portugal, Spain, Bulgaria, and Poland. It would not be surprising if Megalodon teeth were also found in other coastal areas of Europe.

Of course, there are many other such sites worldwide, way too many to list here. But, you should be getting the idea. If you want to find fossils, of any kind, you need to go where those animals lived, and you need to know what that area looked like when they were alive.

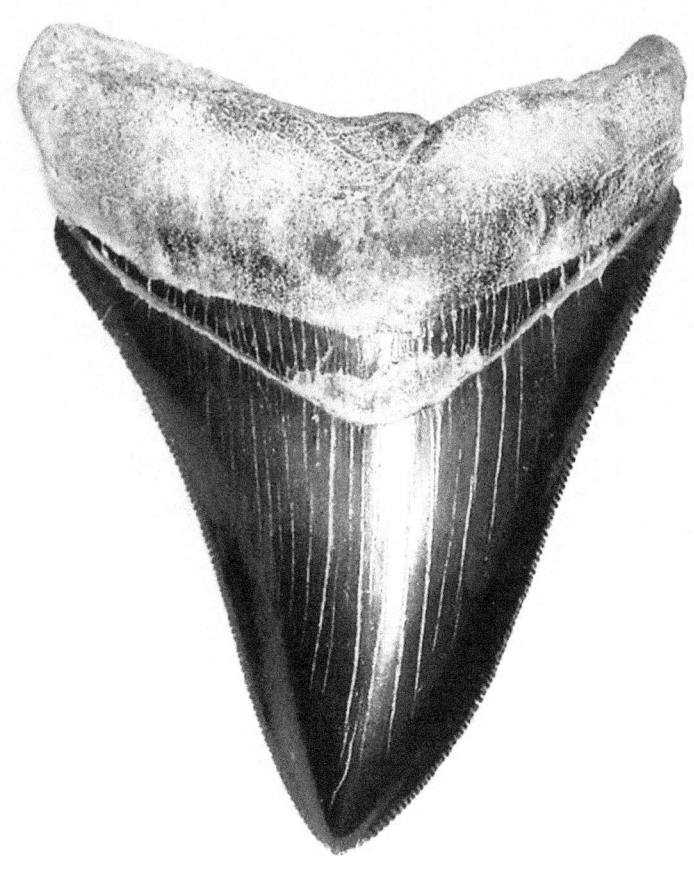

Figure 38. 4.75 inch (12.1 cm) Megalodon tooth.

Figure 39. Photo of the author.

About The Author

Robert L. (Bob) Fuqua was born in south-central Kansas in October 1949 and grew up in a town that never had more than 100 residents. His grade school had one teacher teaching grades one through eight in one room. This school had outhouses and a pitcher pump for water until his third-grade year. One of his favorite childhood memories is finding fossil shark teeth in the chalk bluffs of western Kansas. These chalk bluffs were left behind when the Western Interior Sea subsided.

In 1971, he graduated from Kansas State University with a degree in Mechanical Engineering. He then served in the U.S. Air Force for four years doing intelligence work. That led to a civilian career in intelligence in Maryland.

Along the way he received his SCUBA certification in 1977, and has enjoyed diving in quarries, on coral reefs, and on shipwrecks from New Jersey to Florida. In 1979, he joined the Laurel, Maryland Volunteer Rescue Squad as an Emergency Medical Technician and as a diver in the Marine Division. He participated in numerous search and rescue operations and was also trained in ice diving. In 1985, he was selected as a charter SCUBA diver at the new National Aquarium in Baltimore and eventually became the Sunday Dive Captain. This work included diving with and feeding numerous sharks. After five years, he had to give up that position to take a four-year field assignment in Hawaii.

During his career in intelligence, he served as a

technical analyst, Technical Director, and Chief of numerous organizations. He was an Adjunct Faculty member of the National Cryptologic School (NCS) for many years during his career and taught classes in mathematics and technical analysis. The last few years of his career he was named as the Dean of the Center for Cryptology at the NCS and was responsible for all of the school's intelligence training. He retired in 2002 with over 30 years' service. His wife Linda had previously retired, also with over 30 years' service in intelligence.

They moved to Venice, Florida in 2004 and soon after Bob started diving for fossils and fossil shark teeth in the Gulf of Mexico. He eventually found over 15,000 fossil shark teeth and numerous Megalodon teeth. He had previously written a book on keeping saltwater aquariums in 1995. So, in 2011 he wrote his second book *Hunting Fossil Shark Teeth In Venice, Florida - The Complete Guide: On The Beach, SCUBA, and Inland*. This popular book led to a busy schedule of giving lectures on fossil shark teeth to fossil clubs, SCUBA clubs, civic organizations and a local college. He eventually wrote a total of eight books including this one.

In 2021, Bob and his wife were ready for a change and moved to the east coast of Florida, south of Cape Canaveral, where he has already found several fossil shark teeth. His other interests include astronomy, bicycle riding, stand-up paddle boarding, sailing his Hobie Adventure Island, and driving the 1976 Triumph TR6 sports car he restored.

www.ingramcontent.com/pod-product-compliance
Lightning Source LLC
Chambersburg PA
CBHW050250220526
45465CB00002B/623